RUSSELL

The Story of an Historic Narrow Gauge Steam Locomotive

Andrew Neale

Welsh Highland Light Railway (1964) Ltd.

RUSSELL

The Story of an Historic Narrow Gauge Steam Locomotive

© 1996 Welsh Highland Light Railway (1964) Ltd.

Author:
Andrew Neale.

Book design, Cover Artwork & Drawings by:
Roy C. Link, 1 Station Cottages, Harling Road,
East Harling, Norwich, Norfolk, NR16 2QP.

Printed in Great Britain by:
Amadeus Press Limited, 517 Leeds Road,
Huddersfield, West Yorkshire, HD2 1YJ.

Published by:
Welsh Highland Light Railway (1964) Ltd.

Distributed by:
Cwmni Rheilffordd Beddgelert Cyf.

ISBN 0950 1178 3 8

INTRODUCTION

RUSSELL in 'as built' condition, at Dinas Junction in 1909. (Locomotive Publishing Co.)

The Hunslet 2-6-2 tank locomotive RUSSELL is the only one of the three original Welsh Highland Railway's steam locomotives that still survives today. Before telling this historic locomotive's own life story it is first necessary to say a little about the railway itself.

The Welsh Highland Railway, which opened throughout on 1st. June, 1923 ran for nearly 22 miles from Dinas Junction, 3 miles south of Caernarvon through Beddgelert and the Aberglaslyn Pass to link up with the Festiniog Railway at Portmadoc. The railway was actually a link up between two earlier lines, the North Wales Narrow Gauge Railways Company which opened in stages from Dinas to Rhyd-Ddu (South Snowdon) with a branch to Bryngwyn between 1877 and 1881, and the Portmadoc, Beddgelert & South Snowdon Railway. The latter was the last of several unsuccessful attempts to convert the horse worked Croesor Tramway to a conventional railway and extend it to join the NWNGR at South Snowdon. The Croesor line had been built in 1864 to connect various quarries in the Croesor valley with Portmadoc harbour.

The P.B.S.S.R. was originally intended to be worked by overhead wire electric traction but despite this the steam locomotive that was to become RUSSELL was ordered. On the surface this seems a surprising decision, for construction of the railway was very far from complete. The reason lay in the close liaison with the NWNGR, whose own locomotives were by now needing replacement or heavy repairs. When completed RUSSELL went straight to the NWNGR at Dinas as part of their ordinary stock, and as work on the PBSSR soon faltered away to nothing, there she remained.

Rarely is a steam locomotive a completely new design, invariably either the general concept is "worked up" from previous similar locomotives or at the very least major components such as wheels, motion parts, cylinders, boiler etc. are identical or closely similar to those incorporated in earlier locomotives.

RUSSELL can be said to have had two such locomotive "parents". The first were a class of 2ft. 6in. gauge 2-6-2T for the Sierra Leone Government Railways that were first built in 1898, and the other was LEEDS No. 1, a 2ft. gauge 0-6-2T (Hunslet 865 of 1905) supplied to Leeds Corporation Waterworks. From the the Sierra Leone engines came the coupled wheels, springs, axleboxes, screw reversing gear, coupling and connecting rods, pistons, crossheads and slidebars whilst the boiler and firebox were identical to LEEDS NO.1.

One major difference from the Sierra Leone engines was in the pony truck design. The African engines had plate frame pony trucks with outside axleboxes, helical springs either side of the axleboxes and main frames cut away sufficiently for enough sideways movement of the truck to negotiate a 120 feet radius curve.

RUSSELL, however, had a cast steel "cannon" type pony truck axlebox which embraced the axle between the wheels in one casting, and the suspension was by means of a central helical spring. Cutting away of the main frames was much reduced as the locomotives had merely to negotiate a 198ft. minimum radius curve. This side movement was controlled by swing links and not by pins, a feature more in keeping with standard gauge main line practice than narrow gauge. The reason for its adoption is unknown.

Like LEEDS No. 1, RUSSELL was fitted with Walschaerts valve gear, a feature common to a number of other Hunslet narrow gauge locomotives up to this time, although not employed on a Hunslet-built standard gauge locomotive until 1912.

The specification was altered slightly before RUSSELL was completed: the total wheelbase was reduced two inches to 15ft.6in. and the gauge reduced ¼in. to 1ft. 11¼ in. (The N.W.N.G.Railway. on which RUSSELL was to run was variously recorded as 1ft. 11¼ in., 1ft. 11½ in. and 2ft. as its official gauge!). Extreme height was not to exceed 8ft. 11in. and width 6ft. 5in. Amongst the detail fittings were smokebox spark arrester, Ramsbottom type safety valves, screw reverse, Westinghouse brake, four sandboxes feeding sand to the rails in front of each outer coupled wheel, Westinghouse brake, Wakefield sight feed lubricator coupled to each valve chest with a Furness lubricator in the front end of each cylinder, hand brake and centre buffer with side chains.

The name RUSSELL was bestowed in honour of James Cholmeley Russell, Receiver to the N.W.N.G.

Above: Makers photograph of Hunslet 673-675 of 1898, the initial batch of 2ft 6in gauge 2-6-2 side tanks for Sierra Leone Govt. Railways. Below: Leeds No. 1, Hunslet 865 of 1905, the 2ft gauge 0-6-2 tank for the Leeds Corporation Waterworks railway at Masham.

Rly. Co. Two years later the compliment was returned when the N.W.N.G. line's new Hunslet locomotive was named GOWRIE after Gowrie Colquhoun Aitchison, General Manager of the P.B.S.S.Rly.

Hunslet had sent a specification for Russell on 31st. January 1906, with G.C.Aitchison placing the order on the 13th February. Design and construction were remarkably rapid with the completed locomotive undergoing its steam trials on 26th. May and leaving Leeds for North Wales on the 29th. This rapid construction has prompted several speculative theories such as a suggestion that RUSSELL was adapted from a part built loco intended for use elsewhere but this is definitely not so.

Painting was in N.W.N.G.Railway. livery, red-brown lined black edged yellow either side with bufferbeams and base of brass nameplates painted vermilion.

IN SERVICE IN WALES

RUSSELL and GOWRIE at Dinas Junction in 1909. (The late K.A.C.R. Nunn)

Once delivered RUSSELL settled into regular service on the N.W.N.G. Railway, being joined in 1908 by another new Hunslet locomotive GOWRIE (Works No. 979 of 1908), an 0-6-4 tank of Single Fairlie design. The coming of the First World War accelerated the gradual decline in traffic. Passenger traffic had already ceased on the Bryngwyn branch on 31st. December, 1913 and was steadily falling on the main line. Slate quarries either closed or reduced their output and eventually on 31st. October, 1916 trains were henceforth run purely on an 'as required' basis for freight only. In the circumstances RUSSELL handled most of the traffic, with GOWRIE being sold for Government service and the two Vulcan Fairlie locos cannibalised into one 'good' loco in 1917. This infrequent freight service continued until the incorporation of the Welsh Highland Railway on 30 March, 1922, with powers to acquire the two existing companies and complete the link through Beddgelert to Portmadoc.

The N.W.N.G.Railway and the P.B.S.S.Railway were allotted shares in the new company as purchase money. Capital to complete the construction of the line was subscribed by national Government and various local authorities against the issue of debenture stock. RUSSELL and MOEL TRYFAN, the other surviving N.W.N.G.R. loco, thus passed into Welsh Highland ownership.

The original N.W.N.G.R. route from Dinas to Rhyd-Ddu was reconditioned and reopened to traffic on 31 July, 1922 followed by the rebuilding of the Croesor Tramway from Portmadoc to Croesor Junction. The new 8-¾ mile connecting line from there to Rhyd-Ddu was then completed, opening on 1st. June, 1923, with a connecting line enabling through running to the Festiniog Railway opening a week later on 8th. June. The Bryngwyn branch was never reopened to passenger traffic.

Under the new ownership the Welsh Highland and Festiniog Railways shared both a common chairman,

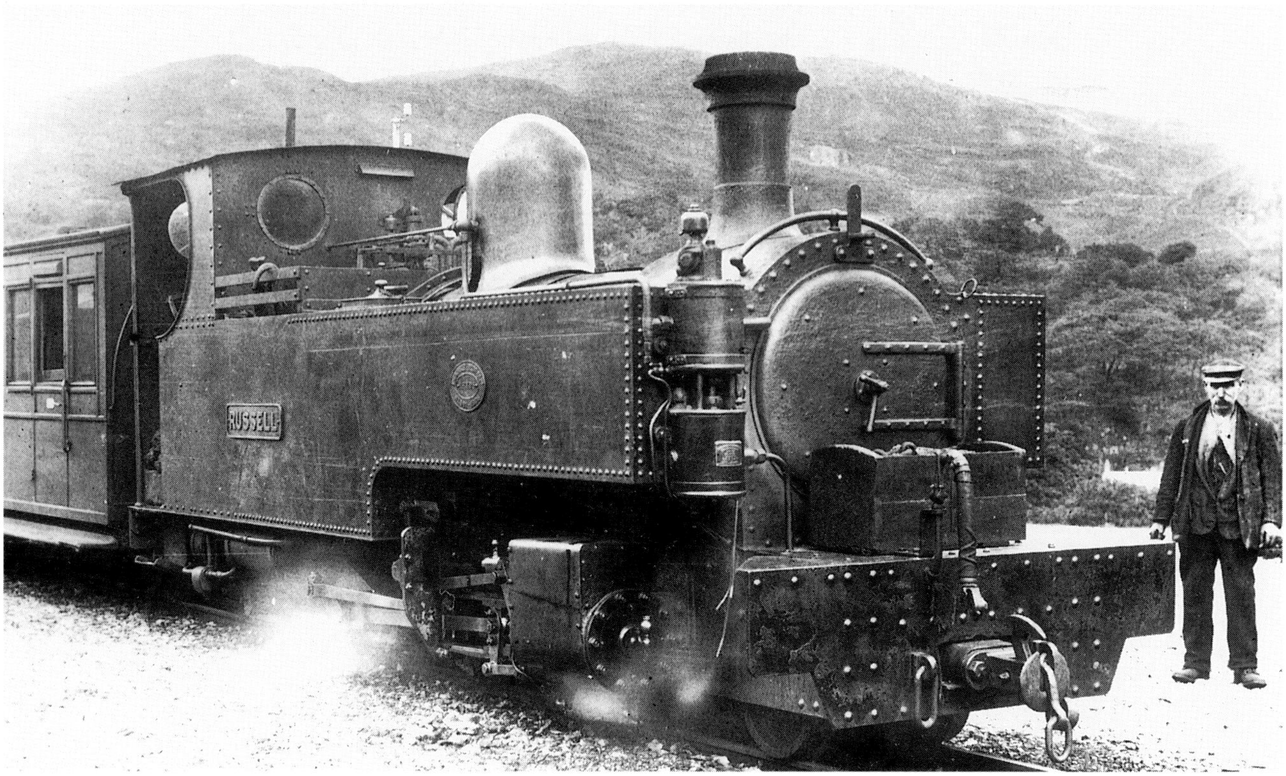

RUSSELL in un cut down condition at Beddgelert in 1923. Driver W. H. Williams stands alongside. (LGRP)

Henry Joseph Jack, Managing Director of the North Wales Power & Traction Co. Ltd.at Dolgarrog and an Engineer/Superintendent, the legendary Colonel H.F.Stephens, who controlled a number of minor railways from his offices at Tonbridge, Kent.

The acute locomotive shortage meant that Festiniog locomotives had to be borrowed to work some services. On the other hand, neither MOEL TRYFAN or RUSSELL could work over the Festiniog due to the severe loading gauge restrictions on that line. On Jack's insistence both locomotives were cut down in height. MOEL TRYFAN was the first to be altered, the cab and chimney being cut down,a vacuum brake substituted for the original Westinghouse one and the locomotive generally overhauled. She was also turned so as to run chimney first into Dinas. After this she became a useful member of the fleet whose daily duties included a Dinas-Blaenau Festiniog return run.

RUSSELL was a different story however. A vacuum brake had already been fitted at Boston Lodge over the 1923 winter and the drastic cutting down now

143, TONBRIDGE.

Your Reference.

FR. I

H. F. STEPHENS, *M.I.C.E*

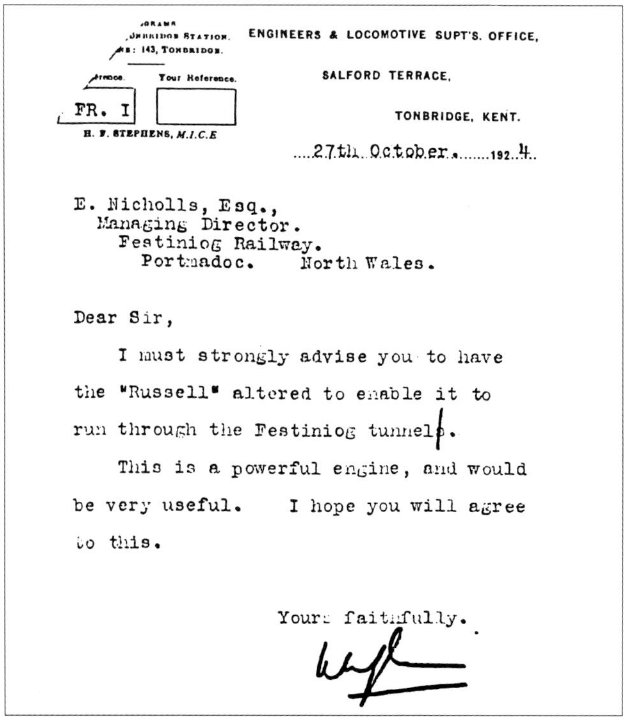

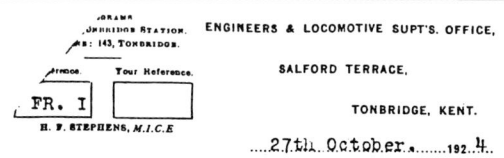

ENGINEERS & LOCOMOTIVE SUPT'S. OFFICE,

SALFORD TERRACE,

TONBRIDGE, KENT.

27th October 1924

E. Nicholls, Esq.,
Managing Director.
Festiniog Railway.
Portmadoc. North Wales.

Dear Sir,

I must strongly advise you to have the "Russell" altered to enable it to run through the Festiniog tunnels.

This is a powerful engine, and would be very useful. I hope you will agree to this.

Yours faithfully.

required was carried out by two Boston Lodge staff at Dinas shed. The chimney cap was removed and a section cut from the dome. The dome cover was then replaced, necessitating a new regulator fitting. The cab was cut down so that it was now only possible to stand up in the centre of it. The upper part of the cab back sheet could now be removed in hot weather. The boiler top sandpot had been removed earlier.

Sadly, not only did these alterations completely spoil RUSSELL's appearance, but they were unsuccessful. On a trial run on the F.R. her tank sides still scraped various rock outcrops in Moelwyn tunnel and in the event of a serious mishap there her crew could not have escaped. Hence forward RUSSELL worked no further from Dinas than Portmadoc Harbour other than the odd visit to Boston Lodge for attention.

A third steam locomotive was obviously needed and one soon arrived on 4th. July, 1923. This was 590, a Baldwin (U.S.A.) 4-6-0T, one of 495 similar locos built for service on the extensive 600mm (1ft. 11⅝ in.) gauge light railways operated by the British Army behind the trenches in World War 1. Too large to be cut down for Festiniog service, and unpopular due to her rough riding characteristics, she was initially confined to Bryngwyn branch traffic but was later used on more general duties.

Traffic on the Welsh Highland Railway was disappointing from the start and steadily declined. RUSSELL and MOEL TRYFAN were the regular locomotives with the Baldwin as spare engine. In 1927 Colonel Stephens was appointed as receiver to the Welsh Highland Railway, a position he retained until his death in 1931.

By the end of the 1929 season RUSSELL was laid off for repairs. Her wheels had been sent back to Hunslet for re-tyring but the makers not unreasonably refused to release them until the work had been paid for. With the W.H.R. quite unable to do so, the impasse was only resolved when Caernarvonshire County Councils stepped in and paid the bill.

Whilst RUSSELL was out of service MOEL TRYFAN operated a passenger service on Mondays, Wednesdays and Fridays only and the Baldwin worked the Bryngwyn branch freight (primarily slate) on Tuesdays and Thursdays. Both locomotives had to be steamed as the Baldwin was considered unsafe for passenger duties at this time and MOEL TRYFAN was not powerful enough for the slate

RUSSELL in newly cut down form is seen at Portmadoc Harbour station in the summer of 1925 (The late H.G.W. Houshold)

RUSSELL and the Baldwin with their respective trains are seen crossing at Beddgelert in 1935. Note that RUSSELL is running with the upper cab back sheet removed. (The late R.H.G. Simpson - Lens of Sutton)

traffic. An extra fitter was taken on at Boston Lodge in June 1931 to complete RUSSELL. Despite this, traffic was now so poor that services were suspended entirely at the close of the 1931 season, although a twice weekly goods service was reinstated on 2nd. November.

To avoid paying charges to the G.W.R. services at Portmadoc now terminated north of the crossing with it. RUSSELL worked the majority of services in 1932 and 1933, the passenger services being reduced to a single return working over a short summer season.

By the end of 1933 it looked as if the railway was finished, although there were thoughts of retaining the Croesor section to facilitate an anticipated reopening of the valley's slate quarries. But at the very last moment, things suddenly changed when the Festiniog Railway announced it was taking a 42 year lease of the line. Passenger services under the new regime commenced on 9th. July, 1934. Once again the passenger service was summer only, ceasing on 13th. October, although a freight service continued throughout the winter.

MOEL TRYFAN had been repaired at Boston Lodge early in 1934 and during that summer RUSSELL and the Baldwin were also overhauled there. As a consequence RUSSELL emerged in a new light green livery and the Baldwin now became a reddish-brown in place of its former black livery. The passenger stock was also repainted in a variety of bright colours.

Sadly, the Festiniog's decision to lease the W.H.R. was a disaster. A loss of £200 was made on the 1934 season and this continued steadily over the next two years, so when the summer passenger service ceased on 26th. September, 1936 it was decided not to reopen the following year.

Over the following months an occasional freight service continued to run between Dinas, South Snowdon and Beddgelert but this failed to even cover its own minimal operating costs and the final train ran

Above: RUSSELL and train pause at Beddgelert. (The late C.R.L.Coles)
Below: On a rather better day than the picture above, the passengers make the most of the rather lengthy stop for refreshments whilst RUSSELL basks in the sun, 8th August 1935. (The late H.F.Wheeler - courtesy R.S.Carpenter)

Above: RUSSELL with a late afternoon return working pauses for water at Beddgelert 1935. (Locomotive Publishing Co.)
Below: Looking well kempt in her new Festiniog livery, RUSSELL poses with her crew before leaving Dinas in 1935. (R.Priestley)

RUSSELL

RUSSELL and 590 are seen lying out of use in the derelict locomotive shed at Dinas Junction shortly before demolition began in 1941. (The late J.F.Bolton)

on 1st. June, 1937. On the 19th. of that month RUSSELL left Dinas Junction early in the morning, collecting all the Festiniog wagons left on the line and delivered them to Portmadoc, where they were left in the sidings behind the harbour. RUSSELL then ran across the Cob to collect the Baldwin from Boston Lodge, propelling her back to Harbour Station. Here the remaining W.H.R. wagons were attached and the train set off for Dinas. More wagons were collected at Beddgelert and the journey continued. However a combination of badly overgrown track and severe curves on the approach to Hafod Ruffydd summit caused RUSSELL to stall. It was necessary to return to Beddgelert and divide the train before making another attempt.

This second attempt was successful and RUSSELL returned to Dinas late in the evening. The following Monday RUSSELL returned to Beddgelert, collected the remaining wagons and returned to Dinas, stopping en route to pick up any odd W.H.R. wagons still remaining in intermediate sidings. Once back at Dinas RUSSELL joined the Baldwin in the locomotive shed, where both were to remain until 1942.

The entire railway then lay derelict until 1941 when the acute war time shortage of steel caused the Ministry of Supply to requisition everything for scrap or further use, under an order dated 13th. March, 1941. The demolition contract was awarded to George Cohen, Sons & Co. Ltd. who began work in August 1941, bringing in two Simplex petrol locomotives to haul the demolition trains.

The Baldwin was considered worthless so was cut up outside the shed in August 1942 but RUSSELL was destined to see further service. The Army had decided to retain a section of the line near Pitts Head for anti-tank gun practice using redundant slate wagons from the Festiniog Railway and had thoughts of retaining RUSSELL for this work. Ultimately this did not happen and RUSSELL was sold by the Ministry of Supply, but not before her name and maker's plates,

Hook Norton quarries. RUSSELL, then still running as a 2-6-2 tank, brings in a train from Redlands quarry on 22nd July 1943. (L.W.Perkins)

plus those of the Baldwin, had been removed and sent to York Museum at the instigation of Mr. V. Boyd-Carpenter. RUSSELL was purchased in May 1942 by the Brymbo Steel Co. Ltd.,who needed more locomotives at their Hook Norton ironstone pits in Oxfordshire to cope with the increased war time demand for ore. En route to Hook Norton RUSSELL was given a thorough overhaul at Brymbo steel works. As a part of this work RUSSELL was repainted in the standard Hook Norton livery, viz grey with red side rods and black lining. New RUSSELL nameplates were cast to replace those removed at Dinas. These had taller letters and cut-away corners and were mounted higher up and further back on the tanks than the originals; once again the background was red.

The extensive 2ft. gauge rail system at Hook Norton dated back to 1898. The first locomotives were an attractive pair of Hudswell Clarke 0-4-2ST named GWEN (built 1899) and JOAN (built 1915), with an ex War Department Light Railways 4-6-0T similar to 590 arriving in May 1919. RUSSELL was followed by a second Hunslet from North Wales in September, 1942. This was BETTY, an 0-4-0ST from the Yr Eifl quarries, Trevor, of the Penmaenmawr & Welsh Granite Co.Ltd. To accommodate RUSSELL a corrugated iron lean-to extension was built over a new line laid in beside the locomotive shed.

Although RUSSELL could easily handle the loads made up in the quarry she was not a success. The track at Hook Norton was typical of such quarry systems, being sharply curved, light and unevenly laid and this led to frequent derailments. In an effort to cure the problem the front pony truck was removed, and even the rear one at one time, but the derailments persisted. Even so, such was the wartime demand for iron ore that both RUSSELL and the 4-6-0T, plus two of the three smaller engines, had to be steamed on most days.

This wartime effort marked a peak in the quarry's output which fell away rapidly at the war's end with

*RUSSELL in service on Fayle's Tramway, Purbeck, Dorset, about 1951.
(C.J. Keylock collection)*

the quarry closing completely in June 1946. After two year's disuse RUSSELL was removed by road in the summer of 1948 to a Ministry sale at Weyhill, near Andover, from where she was purchased by Messrs. B.Fayle & Co. Ltd. for use on their tramway at Purbeck in Dorset.

Fayle's Tramway was a very old system of 3ft. 9in. gauge that had originally been built to carry ball clay from Norden, near Corfe, to wharves at Middlebere and Goathorn. By 1948 it had been reduced to about 4 miles around Norden with clay transhipped to the former LSWR Swanage branch at that point. By then the line's original two locomotives and stock were so worn out that it was decided to relay the line to 2ft. gauge, using standard steel "V-skip" side tippers and a motley collection of secondhand small diesel locomotives as well as RUSSELL.

Unsurprisingly, although RUSSELL originally saw regular service the light quarry track again caused frequent derailments. Once more the attempted remedy was to remove the front truck and run as an 0-6-2T but with no more success than at Hook Norton. By 1953 RUSSELL was in poor mechanical order and when a driving axle sheared that autumn she was withdrawn. When RUSSELL first arrived at Norden she was overhauled and painted a dull maroon but later this was changed to green.

It was at this point that moves were made to save this historic locomotive for posterity. In October, 1953 the Birmingham Locomotive Club were negotiating with Messrs. Pike, Fayle & Co.Ltd., successors to B.Fayle & Co. Ltd., to secure SECUNDUS, another historic narrow gauge locomotive, for preservation. During the talks they were also offered RUSSELL at the scrap price of £70. Subject to finding a suitable home it was decided to accept the offer. Although Caernarvon Town Council were unable to assist, the Talyllyn Railway Preservation Society agreed to accept RUSSELL at their proposed Narrow Gauge Railway Museum at Tywyn and so a preservation fund was duly opened.

A closer inspection of RUSSELL by club members revealed that considerable work would be needed to restore her to appropriate museum condition. It was decided to undertake this work in Birmingham and agreement was reached with the Western Region of

RUSSELL at Tywyn museum on 22nd August 1955, the day of its arrival. (N.J.Allcock)

Goods Depot. However as the Talyllyn R.P.S. expressed a willingness to assist in the restoration work the locomotive was actually moved direct to Tywyn on 22nd. August, 1955. Here she was placed on a short length of track outside the Museum building, adjoining the departure platform at Wharf Station. A tarpaulin was provided for protection during the winter months.

Working parties of B.L.C. members visited Tywyn in March and June 1956 during which the locomotive was cleaned and repainted. The broken axle was removed and sent to Leeds where the Hunslet Engine Co. Ltd. had very generously agreed to repair it free of charge. Unfortunately the continued exposure to the salt laden air and the impossibility of providing covered accommodation caused the B.L.C. to become increasingly concerned at the locomotive's deteriorating condition.

When serious proposals were made to reopen the former Welsh Highland Railway it was natural to consider the long term future of the sole surviving locomotive. After talks between the B.L.C. and the Welsh Highland Light Railway (1964) Ltd., it was agreed that ownership of RUSSELL should pass to the latter provided covered storage was available. This undertaking was willingly given so by mid-1965 RUSSELL was once again owned by a company dedicated to reopening the former W.H.R. Until restoration of the line began, RUSSELL would be stored in Shropshire.

The actual move from Tywyn began on 12th. April, 1965, when the engine was loaded on to a B.R. "Flatrol" wagon prior to removal to Shrewsbury. After a day spent at Machynlleth, RUSSELL arrived at Shrewsbury on April, 14th. A B.R. steam crane transferred her to a Welsh Highland Company's low loader for transport to their workshops on the Kinnerley Junction site of the former Shropshire & Montgomeryshire Light Railway.

RESTORATION AND RETURN TO STEAM

RUSSELL

RUSSELL stands outside the 1964 Welsh Highland Company's workshops at Kinnerley Shropshire on 28th of September 1969. (N.S.D.Evans)

Initially little work was done on RUSSELL other than another cosmetic repaint. The new company had more pressing priorities such as the need to acquire its trackbed and lay track. But problems with the trackbed acquisition prevented any real progress here and thoughts turned to RUSSELL's restoration as a practical way of making progress and maintaining member's morale.

As a first step a "Get RUSSELL steaming" campaign was launched on 20th. September,1968 with the aim of raising around £6,000 to restore the locomotive to running order. A good initial response prompted the decision to order a replacement boiler from the Hunslet Engine Co. at a quoted price of £3,440 and RUSSELL was moved there in November 1969. This was a bold step, for not all the necessary money had yet been raised, but Hunslet very generously agreed to take payment in instalments, and even agreed to the reboiled locomotive being moved to Steamtown,Carnforth, before full payment had been made.

A difficult problem now arose as to priorities. Maximum effort was now being given to the acquisition of the "Beddgelert Sidings" area in Portmadoc as a terminus for the new railway,leaving nothing to fund further restoration work at Steamtown or even complete the final payment on the boiler. But fund raising continued and the July 1973 W.H.R. Journal was able to carry the proud announcement that RUSSELL had been re-boilered and the work fully paid for.

Further work on RUSSELL could at last be considered but sadly spiralling prices meant that the cost of restoration at Carnforth was now far in excess of that envisaged five years earlier. The alternative was

Continued on page 18...

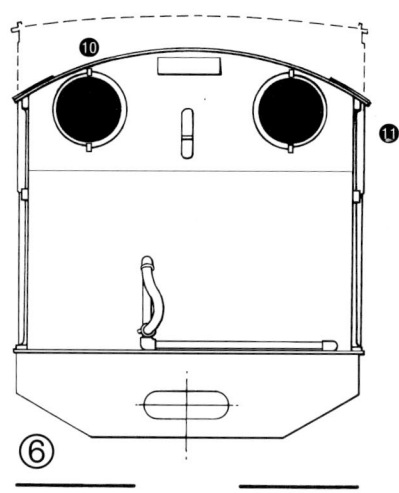
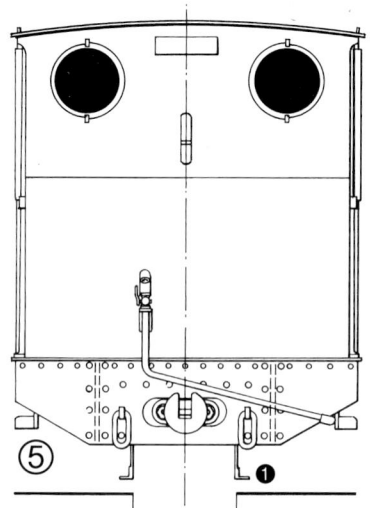
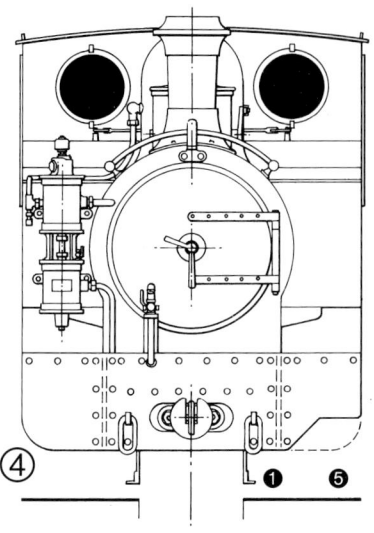

W.H.R. 'RUSSELL'

Built by the Hunslet Engine Company Limited, Leeds. Works Number 901 of 1906.

©1996 W.H.L.R.Ltd
Drawn by Roy C. Link.   Scale 1:43·5

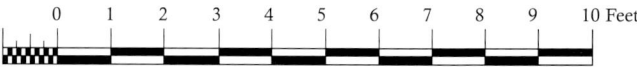

NOTES

A variety of material has been referred to while producing these drawings. Makers drawings and photographs were the main sources. The following notes are intended as a guide to the many changes made to 'RUSSELL' over the years. Every attempt has been made to verify these, but ultimately it is strongly advised that reference to period photographs are made, particulary if a model is being built. Only major bolt head detail is drawn, for clarity. Study photographs for rivet placement - which in itself varies depending on period.

① Side elevation of RUSSELL as built.
② Plan view, split along centre line showing RUSSELL as built (lower) and as modified in 1923 (upper).
③ Side elevation of RUSSELL as modified in 1923.*
④ Front elevation of RUSSELL as built.
⑤ Rear elevation of RUSSELL as built.
⑥ Rear elevation of RUSSELL as modified in 1923.*

* Original outline shown as dashed line. Only altered details shown.

❶ Plate metal guard irons only visible in earliest of photographs - it is assumed they were removed at an early stage.
❷ Sandbox - removed prior to cutting down.
❸ Westinghouse brake piping only shown on plan view.
❹ Westinghouse brake hoses omitted - details unknown.
❺ Buffer beams modified prior to rebuild - possibly at the same time as the sand dome was removed.
❻ Cut-out in lowered cab roof - 1923 rebuild.
❼ Vacuum brake gear and piping - 1923 rebuild.
❽ Piping revised when vacuum brake gear fitted - 1923 rebuild.
❾ Vacuum brake pipe on cab floor - 1923 rebuild.
❿ Cab spectacles lowered front and rear - 1923 rebuild.
⓫ Cab rear upper sheet made removeable - post 1923 rebuild. Photos taken during summer months often show RUSSELL running without the top rear sheet.
⓬ Wheels, cylinders and motion etc., ommited to show frame outline and cut outs.

RUSSELL

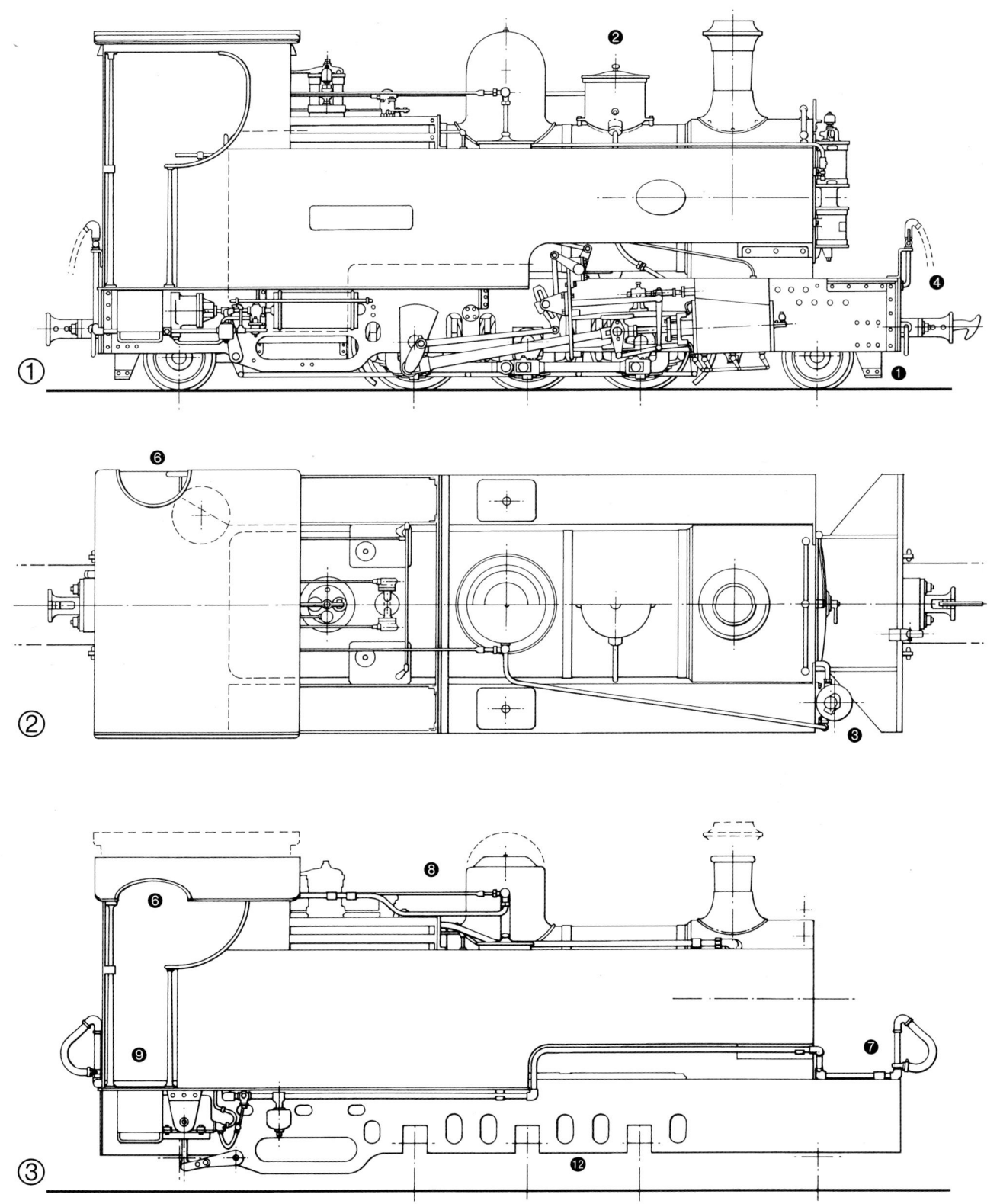

A view of the Hunslet Engine Co. Errecting Shop in November 1969 shortly after RUSSELL had arrived there for repair. In the foreground is one of the two bogies from the Festiniog Railway's 'Fairlie' locomotive MERDDIN EMRYS which had been sent to Hunslet for the fitting of new cylinders. (Hunslet Engine Co.)

---

to undertake restoration using W.H.R. volunteers at a suitable covered site. Messrs. Hills & Bailey Ltd. of Llanberis generously offered to accommodate the locomotive for an annual rental of £50, which would be waived if they carried out work on the locomotive to a higher value during the year. They also agreed to work being undertaken by volunteers or outside contractors.

For the next six years restoration made slow progress at Llanberis. During the period at Llanberis, RUSSELL was stripped right down to component parts, and the cylinders were removed and rebored at the BR marine workshops at Holyhead. The years of operational neglect had resulted in serious wear and damage to the chassis. The right hand axleboxes were seized solid, and had to be removed by drifting out with wedges and a 14lb sledge hammer. A combination of corrosion and wear to all the bearing surfaces dictated that a comprehensive overhaul would be necessary, in addition, the 'U' shaped cast iron horn block castings were irreparably damaged as a result of the locomotive having been operated for many years without horn stays fitted.

By now the land at Beddgelert Sidings and the adjoining Gelert's Farm had been purchased and all efforts needed to be concentrated there so once again work on RUSSELL largely stopped. The dismantled locomotive was moved to Gelert's Farm on 28/29

RUSSELL's frames stand amongst much other glorious clutter in the locomotive shed at Gelert's Farm, in February 1986. (D.W.Allan)

August 1977, the boiler and other parts stored under cover and the frames left on the headshunt.

A fresh start was made in 1980 when the "Friends of RUSSELL Society" was formed to raise the necessary funds for the completion of the work. Membership at an annual subscription of £4.00 was initially confined to the W.H.R. members, but by 1984 sufficient money had been raised to allow work to be resumed. By now the railway had established a comprehensive workshop facility, and a volunteer workforce and management structure, capable of carrying out the extensive overhaul of the locomotive, and this capability enabled most work to be tackled 'in house', supplemented where appropriate with specialist sub contractors under the company's own management. In order to focus the energies harnessed in the workforce, a target was set for having RUSSELL steaming in 1986, the locomotive's 80th birthday. The restoration team was established as part of the company Mechanical Engineering Department, and it was this team who were able to see things through to completion. Fund raising now gathered pace, one particularly successful initiative being the sale of seats on the first four inaugural trains, at prices ranging from £40 to £12. It is pleasing to record that from this point on work was never to be delayed by lack of funds.

Initially, six new driving wheel tyres were ordered from the British Steel Corporation at a cost of

Another view at Gelert's Farm in February 1986, RUSSELL's new boiler is temporarily supported by two four wheel skip underframes awaiting reunification with the frames. (D.W.Allan)

£1101.91 and Bootham Engineering PLC contracted to rectify a bent axle and misaligned flycrank at a cost of £675.00. Further work was undertaken by a local engineering firm, Winson Engineering, who carried out the machining of driving wheel sets and tyres prior to assembly. A new ashpan was made free of charge by Nigel Massey Engineering, material for a new pony truck axle was also donated, and a Yorkshire Area member donated new castings for piston heads and hornblocks, the latter being cast from the original pattern which had been obtained from Hunslet Holdings.

Steady progress was made throughout 1985. During March and April, the frames were needle gunned, shot blasted and primer painted, whilst Winson Engineering undertook the machining of axleboxes and hornblocks in the workshops at Gelert's Farm. The securing of the new hornblocks with new fitted bolts was completed, together with machining and fitting of liner plates and hornstays in readiness for the re-wheeling of the frame. With the driving wheels re-tyred, attention was turned to reclamation of the bearing surfaces of the axleboxes, by machining and fitting of thrust face liners. Overhaul of the springs and manufacture of new adjustable spring hangers, linkage and compensating beams was completed in preparation for re-wheeling. Both pony

Two months later, Colin Blackwell and Mike Fairburn are carefully fitting the new tyres onto RUSSELL's wheels in the main carriage shed at Gelert's Farm. (D.W.Allan)

trucks were completely overhauled and fitted with new swing linkages. A replacement leading frame stretcher was fabricated to replace the missing original. The absence of this component from the dismantled kit of parts, led the restoration team to suspect that breakage of the cast iron item in the past was the true reason for the locomotive being operated as an 0-6-2 in the latter years of its industrial career, though with the front truck fitted the lack of vertical suspension compliance over the total wheelbase would render the locomotive very liable to derail on track with poor vertical alignment. Work commenced upon the renovation of the motion, by the removal of many years of grime, by steam cleaning and shot blasting. Most of the parts of the motion were extant, but some bearings were missing and all bearing surfaces were badly worn. In addition, the combination levers were bent, this probably having been done by some fitter possessing more confidence than understanding, particularly since the bending had the effect of making the valve setting worse than prior to this 'ham fisted' modification. The first real milestone in transforming RUSSELL from a kit of parts to complete locomotive came during the last weekend in April, 1986 when the frame was lowered back onto the wheels. The following Tuesday, the boiler was re-united with the frame the necessary

Chief Mechanical Engineer, Mike Fairburn, cracks open the regulator as a rather naked RUSSELL moves under her own steam for the first time in forty five years, 6th December 1986. (D.W.Allan)

crane being provided free of charge by Tom Buckley. The assembly of the slide bars, crossheads and motion continued throughout 1986. In late November, the brass dome cover was returned, having been restored to its original height completely free off charge by Eckold Ltd and Laroc Engineering, both of Tile Hill, Coventry. New cylinder drain cocks were supplied by Guest and Crimes of Rotherham, and all of the boiler fittings were overhauled, either at Gelerts Farm, or by specialists off site, and a new blast pipe assembly was fabricated.

But time was rapidly running out if the goal of steaming RUSSELL in 1986 was to be achieved and many members were convinced that the idea was impossible. To their surprise the January 1987 W.H.R. journal contained a first ever supplement. Just as the magazine's editor, Dave Rowbotham, was about to go to press the Works Manager, Stuart Weatherby, rang with the wonderful news "They steamed RUSSELL yesterday". The journal supplement consisted of an illustrated account of the historic first steaming on Saturday, 6th. December, 1986, which is reproduced verbatim below:

"There was an air of expectancy at Gelerts Farm.

The late Ken Dicks then Company Chairman, carefully checks RUSSELL's progress as she moves onto the main line for the first time. Note temporary 'litter bin' water tank, 6th December 1986. (D.W.Allan)

The day had dawned crisp and clear with the low winter sun adding its own version of warmth to the scene, but this was no ordinary day.

RUSSELL stood just outside the loco shed, festooned with tubes, rather like a patient that had undergone a serious operation, a wisp of smoke escaping from her chimney. Today was to be the climax of ten year's hard work, most of which seemed to have been crammed into the last twelve months.

An hour later and smoke was beginning to billow from her chimney, steam was escaping from valves which had not been troubled with this element for over thirty years. The engineers were fussing like hens round an overgrown chick, testing valves, tending the fire and examining the gauges.

The pressure was starting to rise, time to test the whistle, a deep throated melodious sound echoed back from the Tremadoc sea cliffs, the last time they had resounded to that note was on a flat dismal day, fifty years ago, when she bade them farewell.

The tension as well as the pressure was beginning to mount. The injectors were tested and worked. Everything seemed in order. There was 150lbs on the clock - it was the moment of truth. Mike Fairburn

Success at last! RUSSELL's steam pressure gauge records the first few pounds of steam to be raised in her new boiler, 6th December 1986. (D.W.Allan)

By Febrary 1987 RUSSELL was starting to look like a real locomotive again. By now the newly fabricated side tanks and cab had been fitted but there is still a tremendous amount of work to be done if the lcomotive is to be finished in time for the scheduled ceremony two months later. (D.W.Allan)

cracked the regulator open, the steam hammered into the cylinders and she moved a couple of yards up the track then a couple of yards back - elation! It was not yet time to relax, would she go further? How would the track react? Would she go through the points? It was now or never. Mike on the footplate was looking like a jockey on an unbroken colt, he gave a blast on the whistle and she was off, swathed in smoke and steam, past the water tower, wheels spinning vigorously through the first set of points, then she eased to a halt just short of the new turnouts.

Worried concentration on the General Manager's face as he signals her through the points, relief, she behaves perfectly. Now, at last, on the main line she pauses whilst coal from the Hudson bogie wagon parked conveniently in the headshunt is shovelled onto her footplate. Then back through the points and down to the station at Portmadoc. A moment of concern as she grinds round the curve into the station, it will have to be eased a little. Now, with more confidence, return up the main line to halt at Gelerts Farm groundframe.

The large blue barrel, prominently marked "litter", which is doubling as a temporary water tank, is almost empty. After an abortive attempt to refill it with a bucket, a length of hose is run out from the nearest tap. Tanks replenished, she is moved gently forward

First day in service, Easter Saturday 18th April 1987. RUSSELL poses at the station platform at Porthmadog prior to hauling the inaugural passenger train. At this stage the locomotive was still in unlined works grey livery and sundry items such as cylinder lagging and coal rails had still to be fitted. (D.W.Allan)

beyond the yard turnout, where she pauses whilst a "tender", consisting of a weed killing tank filled with water mounted on a slate wagon, is attached. Whistle blasts, regulator opened and away she goes romping like a puppy up the main line to Pen-y-Mount. The beat of the exhaust brings a lump to the throat - RUSSELL is back home and clearly happy to be so."

Having managed to keep within the promised time for a first steaming, there was still a great deal to do to achieve the planned date of hauling the first passenger train on Easter Saturday, 18th. April, 1987.

A major task was the construction of new side tanks. These were built on a bedplate of rolled steel joists welded to a section of track in the carriage shed The side plates were first assembled and ribs, to which baffle plates were bolted, welded to their inside. The complete assembly was then seam welded inside and out. This was a most unpleasant job, ably carried out by Mike Fairburn, Dave Ruston and Colin Blackwell, who had not only to try and arc weld within the confines of the tank, but endure all the fumes and heat generated as well.

By the end of February the completed tanks were in place and the cab sheets assembled and mounted.

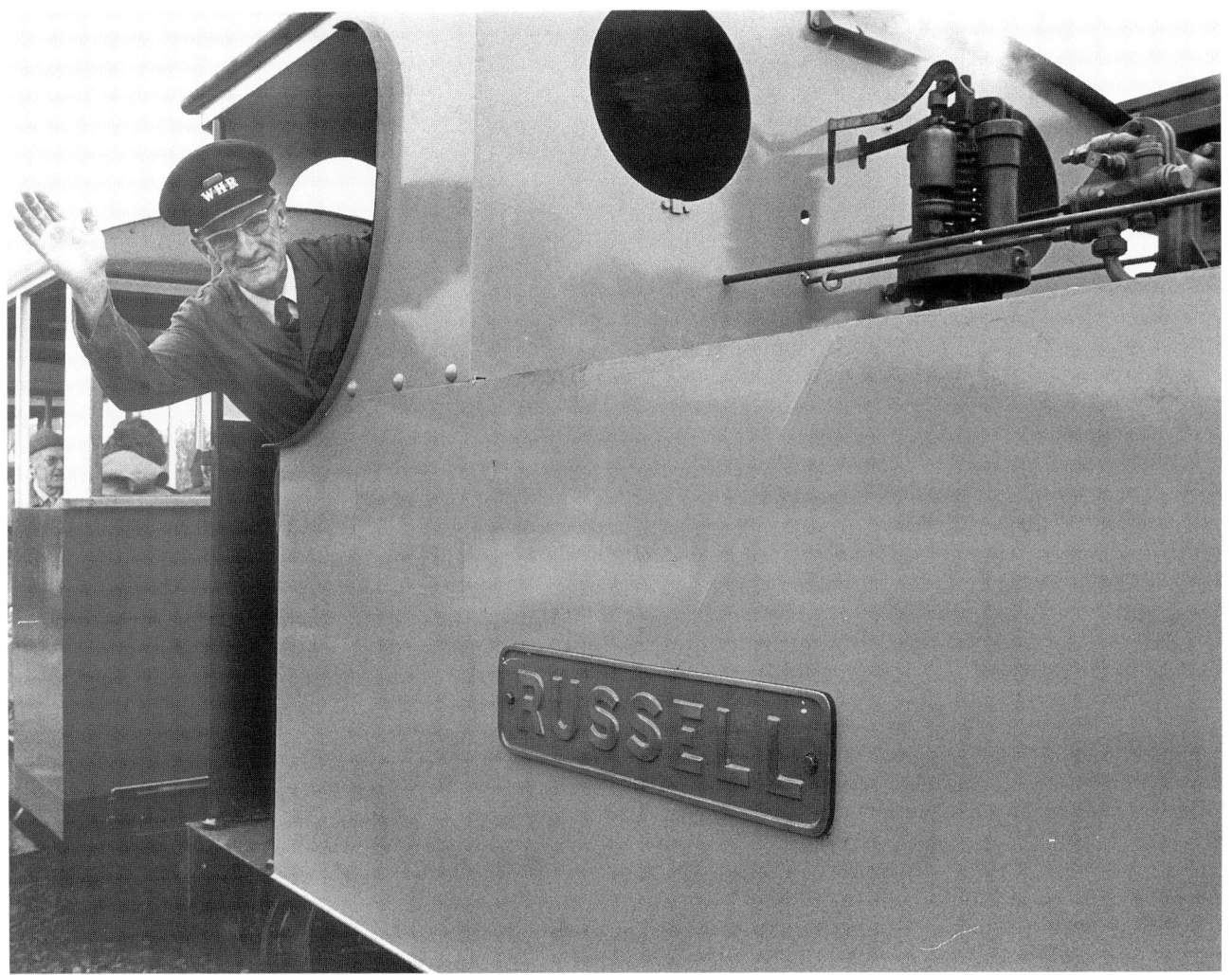

A pleasant feature of RUSSELL's 'first passenger train' ceremony, was the reuniting of the locomotive with its former driver from original Welsh Highland Railway days. Goronwy Roberts proudly poses on the footplate of his former charge.
(D.W.Allan)

Next the boiler was lagged and the Westinghouse pump and associated pipework were fitted. Complete air braking and numerous other small jobs had all to be done. On Friday the 17th of April there were still a team of volunteers making last minute adjustments. Mike and Colin were working until gone midnight, but still managed to be up at 4-30am next morning to light RUSSELL's fire. Once again we quote from David Allan's article in the Journal...

'Easter Saturday arrived, cool and overcast, but this was not going to deter RUSSELL's fans, and well before the due time crowds were turning up in their droves. The car park was soon full, and overflow parking arranged in the former BR goods yard. Platform tickets were being used and a temporary wooden ticket office installed on the bridge over "Y Cyt", efficiently managed by Jim Leyland. Cliff Jones had been appointed Station Master and a resplendent figure he cut - ramrod straight back, immaculate uniform with gleaming buttons, complete with carefully consulted fob-watch. Nothing was going wrong if he had anything to do with it!

At 9-45 a.m. KAREN appeared as light engine

Throughout the 1987 season RUSSELL worked trains still carrying the shop grey livery. On a sunny summer's day in July 1987 she is seen in service with a train running with upper cab backsheet removed as in pre-war days. (D.W.Allan)

manned by Steve Seale and Don Little, but for once KAREN was not the star, but ran into the loop where she waited, simmering expectantly. At 9-55 a.m. the coaches appeared with GLASLYN backing them down under the control of the immaculately turned out guard, Andy Blackwell. At 10 a.m. Myfanwy Roberts officially opened "Russell's Cafe". Catering Manager John Holmes presented her with not only a large bunch of flowers but also a very large kiss. So overcome was he by this event that he declared free tea would be available for everyone all day.

It was now 10.10 a.m. A haunting whistle echoed from the direction of Gelert's Farm. The large crowd thronging both platforms turned to gaze expectantly up the line, and there, appearing out of the grey Porthmadog mist was that familiar outline of a hundred photographs. Wreathed in steam and whistling to acknowledge the salutes from the crowd, RUSSELL in works grey livery slipped gracefully into the station in front of her rake of coaches.

With press and T.V. cameras everywhere there was hardly room to move on the platform but a space was soon made for Company Chairman Ken Dicks. Ken introduced RUSSELL's former driver on the original W.H.R., Goronwy Roberts, who had been invited to re-commission his old loco. It was a poignant moment

Finished at last, the fully restored RUSSELL shunts stock in July 1993.
(D.W.Allan)

– fifty years had passed since the two had last worked together and understandably it was with some emotion that Goronwy unveiled the original Hunslet nameplate and simply said "Welcome home RUSSELL".

The ceremony over, the coaches filled up with those generous "Friends of RUSSELL" who had each contributed forty pounds for the privilege of riding on the first train hauled by the restored RUSSELL. Goronwy gave a footplate interview to BBC Wales reporter Alan Bareham whilst the loco crew made last minute checks and adjustments. Driver Mike Fairburn had been persuaded to wear a white shirt and bow tie whilst fireman Colin Blackwell's dungarees were beautifully pressed.

The green flag was waved, and with a responsive toot the world's most famous narrow gauge steam locomotive was back in service. To cheers and waves from passengers and watching crowds, and a whistle from KAREN, RUSSELL and train moved slowly away. Clattering away over the points and onto the single line, there was a final acknowledging whistle blast before rounding Gelerts Farm curve and disappearing from sight, only a plume of steam marking the train's progress. It was an historic moment and as a prominent member of the railway press commented: "If your Company can achieve this,

Another view of the fully restored RUSSELL – in service July 1993. (D.W. Allan)

then there's nothing beyond your capabilities" – a prophetic thought to end a significant day."

RUSSELL remained in its works grey livery for the rest of the 1987 season but was in original N.W.N.G.R. colours during Easter 1988, the cost being met by a member from south-east England. Over the 1987/8 winter the locomotive was given a very thorough check over and service and any necessary adjustments and remedial work carried out. A vacuum brake was also fitted. On Saturday, 7th. May, 1988 the locomotive took part in a series of events celebrating the 125th. anniversary of steam haulage on the Festiniog, working trains up to Rhiw Goch loop.

Since 1988 RUSSELL has become a regular performer in the W.H.R. locomotive fleet. Over the last few years work has continued on the completion of the restoration, such as the refitting of the original boiler top sandpot. As RUSSELL approaches her 90th birthday in 1996 she can look forward to many years more service, hopefully ultimately on a fully reopened Welsh Highland Railway.

The excellence of the work undertaken in restoring RUSSELL was acknowledged in 1990 by the presentation of the Scania Transport Trust Finalist Award.

Appendix "A"
Principal Dimensions (as built)

The specification of 31 January, 1906 was slightly amended in two parts before "Russell" was completed: the planned wheelbase of 15 ft. 8 in. was reduced "as much as possible," and became 15 ft. 6 in., while the gauge was altered from 1 ft. 11⅞ in. to 1 ft. 11¼ in. Extreme height was not to be more than 8 ft. 11 in. and width 6ft. 5 in.

Additional fittings included spring gear compensating beams, on each side between the coupled wheels which were inside the frames, while the axles were fitted with outside bearings. Other fittings included a spark arrester in the firebox, and ash-pan with side cleaning-out doors, Ramsbottom type safety valves, screw reverse, four sandboxes conducting sand to the rails in front of the coupled wheels at each end, a Wakefield sight-feed lubricator coupled to each valve chest, a Furness lubricator in the front end of each cylinder draincocks in each end of the cylinder Westinghouse brake, hand brake and centre buffer with side chains.

Make and Number	Hunslet Engine Co. Ltd., 901 of 1906
Gauge	1 ft. 11¼ in.
Wheel arrangement	2-6-2T
Cylinders	10¾ in. x 15in.
Tubes (Brass)	97 of 1⅝ in.
Working Pressure	160 Ib./sq. in.
Tractive effort	7425 Ib. (75% w.p.)
Water capacity	440 gallons
Coal capacity	29 cu. ft. (14½ cwt.)
Heating surface- Tubes	345 sq. ft
Heating Surface -Firebox	36 sq. ft.
Grate area	6¼ sq. ft
Coupled wheels (dia.)	2 ft. 4 in
Bogie wheels (dia.)	1 ft. 6in
Coupled wheelbase	5 ft. 6 in
Total wheelbase	15 ft. 6 in
Boiler barrel (length)	8 ft. 1½ in.
Boiler barrel (dia.)	3ft. 1½ in.
Weight (empty)	16 tons
Weight (working order)	20 tons

Appendix "B"
Comparative sizes of well-known Two-Foot Gauge Engines

Engine	Cylinders	Wheel dia.	Pressure	Tractive Effort
V. O.R. ENGINES	11 x 17	30 in.	165 lbs.	10,510 lbs.
RUSSELL	**10¾ x 15**	**28 in.**	**160 lbs.**	**7,425 lbs.**
LYN (Scrapped)	10 x 16	33 in.	180 lbs.	7,418 lbs.
LEW (Scrapped)	10½ x 16	33 in.	160 lbs.	7,269 lbs.
EARL OF MERIONETH	9 x 14	33¼ in	160 lbs.	5,078 lbs
LINDA	10½ x 12	31 in.	140 lbs.	5,078 lbs.
PRINCE	8 x 12	27 in.	140 lbs.	4,489 lbs.

SELECT BIBLIOGRAPHY

NARROW GAUGE RAILWAYS IN NORTH WALES
Charles E. Lee, Rly. Publishing Co., 1945.

NARROW GAUGE RAILS TO PORTMADOC
James I.C. Boyd, Oakwood Press, 1949.

THE WELSH HIGHLAND RAILWAY
Charles E. Lee, David & Charles Ltd. & Welsh Highland Light Railway (1964) Ltd., 1968.

MORE ABOUT THE WELSH HIGHLAND RAILWAY
David & Charles Ltd. & Welsh Highland Light Railway (1964) Ltd., 1966.

INDUSTRIAL RAILWAY RECORD No. 2
Industrial Railway Society, 1963.

INTRODUCING RUSSELL
Peter Deegan, Russell restoration Fund, 1969.

RUSSELL, THE STORY OF A LOCOMOTIVE
Alun Turner, Welsh Highland Light Railway (1964) Ltd., 1990.

NARROW GAUGE RAILWAYS IN SOUTH CAERNARVONSHIRE,
James I.C. Boyd, Oakwood Press, 1972

ACKNOWLEDGEMENTS

This present work is essentially an updated and revised version of two earlier books mentioned in the bibliography above — INTRODUCING RUSSELL and RUSSELL, THE STORY OF A LOCOMOTIVE and due acknowledgement must first be made to Peter Deegan and Alun Turner, the respective authors. Further help has also been given by Messrs. Mike Fairburn, Neil Evans, Malcolm Hindes, Les Blackwell, David Allen and John Keylock for much help with information and illustrations. Last but absolutely by no means least we must all be grateful to those many individuals and organisations, some mentioned within the pages of this little book but many who are not, who freely gave of their time, skills and money to successfully complete the restoration of RUSSELL, a truly historic narrow gauge locomotive.